About Us

Hi! My name is Malia, I'm one of two people behind T-Bot, the most curious Wondroid. My creative husband, Johnathan, created this lovable tiny robot back in 2010 when he was only 15 years old! A decade later, in 2021, we were ready to share T-Bot with the world. Our mission is to re-spark exploration of the world through the lens of wonder.

With T-Bot and the Wondroids, we strive to create opportunities where kids can learn and develop at their own pace. We believe that our activity workbooks are helpful tools for parents who want their children to learn and grow with confidence in themselves.

I'm so proud of what we have achieved thus far. I truly believe you and your kid will fall in love with T-Bot just as I did when my husband first showed the tiny bot to me!

If you enjoy our books and want to help us share our mission, please leave a review – we'd really appreciate it! Thanks for being here!

Luminis Creo LLC

AF887976

Early Math

Luminis Creo LLC books may be purchased for educational, business, or sales promotional use. for information, please email the inquiries at office@luminiscreo.com

First Edition
ISBN: 978-1-960423-28-3

This astonishing book belongs to:

T-Bot wants to give you this very astonishing award!

psssst printable color version :)

Free Goodies?

Scan this QR Code

Section 1: Handwriting 0-10

Instructions:

1. Follow the doted lines.
2. Trace each number.
3. Keep practicing in the blank spaces!
4. Your number writing skills are awesome!

Example: Filled Out

Practice tracing the 0s.
Practice writing the number zero.

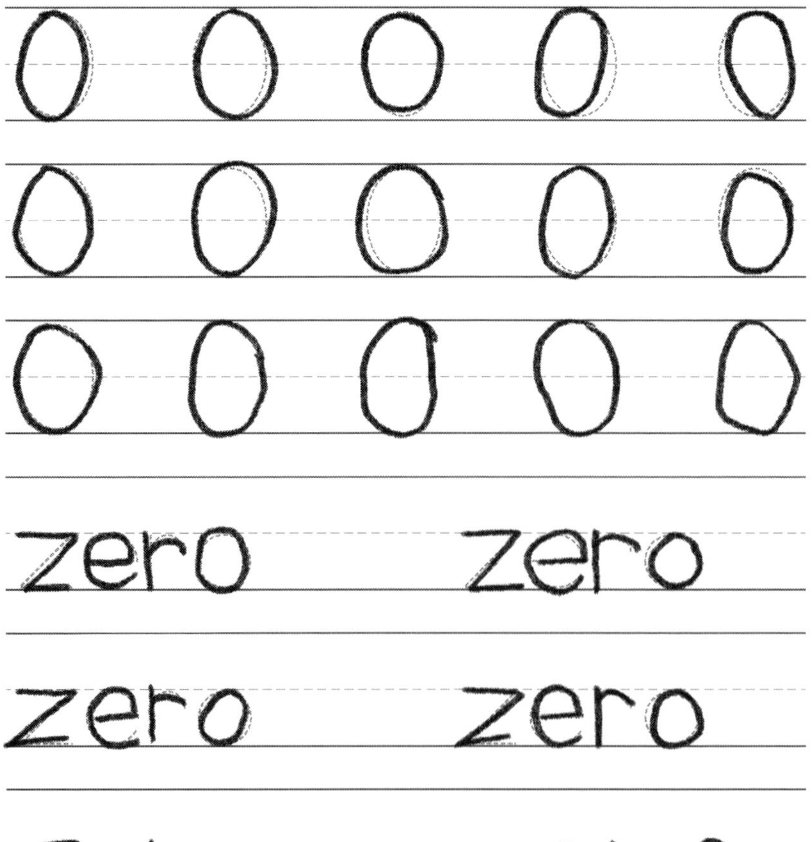

Practice tracing the 0s.
Practice writing the number zero.

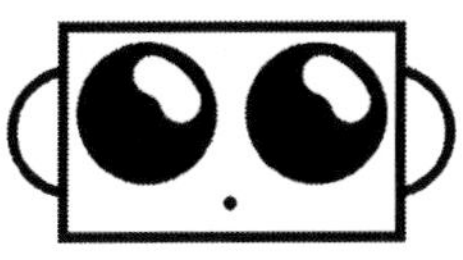
admirable

0 0 0 0 0

0 0 0 0 0

0

zero zero

zero zero

zero

Practice tracing the 1s.
Practice writing the number one.

excellent

1 1 1 1 1

1 1 1 1 1

1

one one

one one

one

Practice tracing the 2s.
Practice writing the number 3.

2 2 2 2 2

2 2 2 2 2

2

two two

two two

two

Practice tracing the 3s.
Practice writing the number 3.

delightful

3 3 3 3 3

3 3 3 3 3

3

three three

three three

three

Practice tracing the 4s.
Practice writing the number 4.

remarkable

4 4 4 4 4

4 4 4 4 4

4

four four

four four

four

Practice tracing the 5s.
Practice writing the number 5.

5 5 5 5 5

5 5 5 5 5

5

five five

five five

five

Practice tracing the 6s.
Practice writing the number 6.

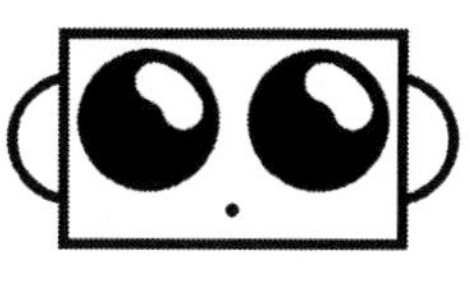

6 6 6 6 6

6 6 6 6 6

6

six six

six six

six

Practice tracing the 7s.
Practice writing the number 7.

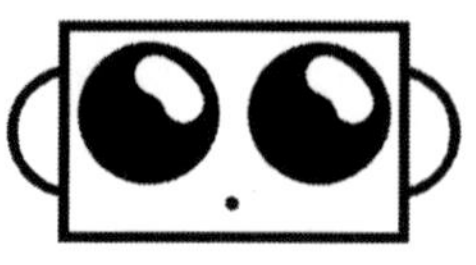

superb

7 7 7 7 7

7 7 7 7 7

7

seven seven

seven seven

seven

Practice tracing the 8s.
Practice writing the number 8.

brilliant

8 8 8 8 8

8 8 8 8 8

8

eight eight

eight eight

eight

Practice tracing the 9s.
Practice writing the number 9.

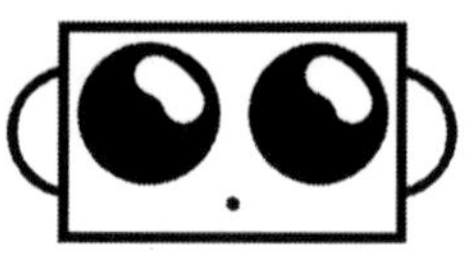

wonderful

9 9 9 9 9

9 9 9 9 9

9

nine nine

nine nine

nine

Practice tracing the 10s.
Practice writing the number 10.

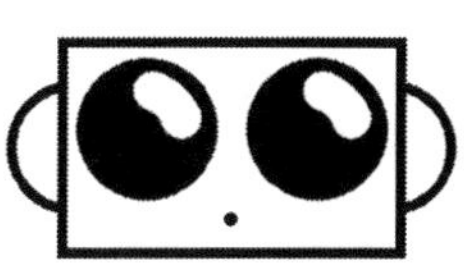
divine

10 10 10 10

10 10 10 10

10

ten ten

ten ten

ten

Section 2:
Find the Number & Connect the Number

Instructions:

1. Look at the objects on the left.
2. Draw a line to the matching number.
3. YAY! Such talent!

Instructions:

1. Read the instructions on each page.
2. Circle all of the correct numbers.
3. YAY! Way to go!

Example: Filled Out

Example: Filled Out

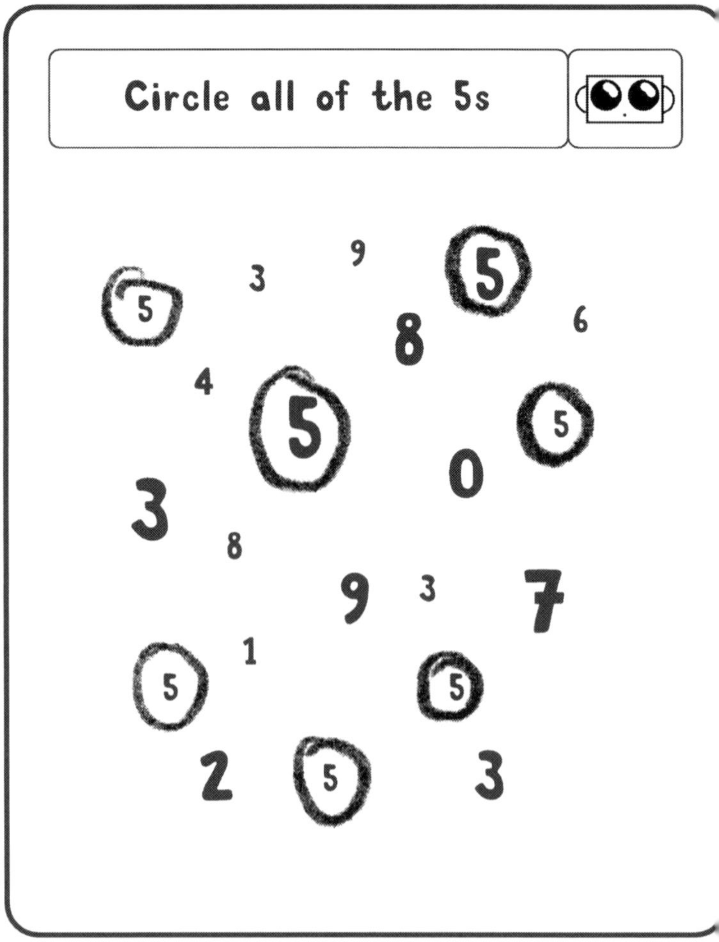

Draw a line to show how many.

9

3

6

1

Circle all of the 3s

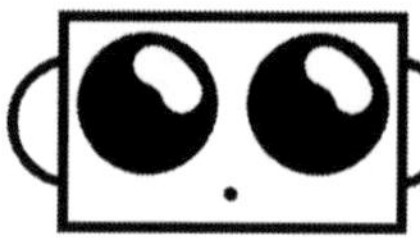
terrific

9 3 3

2 8 6

4

4 3

0

3

8

9 3 7

1

3 5

2 5 3

Draw a line to show how many.

super

4

9

5

7

Circle all of the 4s

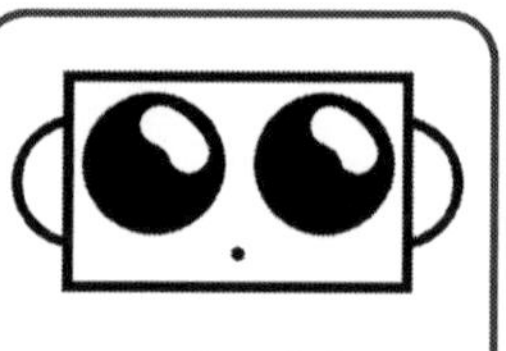

awe-inspiring

3 9 3

2 4 4

4

4 3

0

3

8

9 3 4

4

3 4

4 5 3

Draw a line to show how many.

sensational

8

6

2

1

Circle all of the 1s

wonderful

3 1 3

1 8 1

4

4 3

1 1

8

1 3 7

1

3 5

2 5 1

Draw a line to show how many.

dynamite

8

3

7

2

Circle all of the 7s

astonishing

9

7 3

2

8 6

4

7 3

0

3

7

9 3 7

7

3 5

2 5 7

Draw a line to show how many.

wonderful

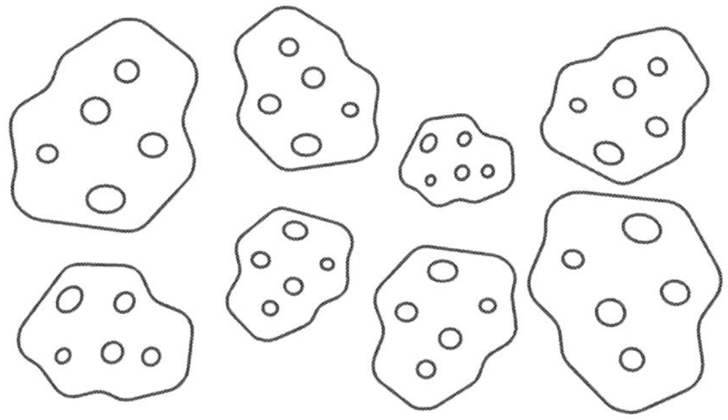

2

5

4

8

Circle all of the 2s

3 9 3

2 8 2

4

2 3

2 2

8

9 3 7

1

3 5

2 5 3

Draw a line to show how many.

astounding

8

3

7

1

Circle all of the 8s

fabulous

9 3 3

2 8 8

8

4 3

0

3

8

9 8 7

8

3 5

2 5 8

Draw a line to show how many.

outstanding

5

4

6

9

Circle all of the 6s

groovy

6 3 3

2 8 6

4 4 3

0

6 8

6 3 7

1

3 5

6 5 6

Draw a line to show how many.

swell

9

3

6

5

Circle all of the 5s

miraculous

9

3 5

5 8 6

4

5 5

0

3

8

9 3 7

1

5 5

2 5 3

Section 3: Circle the Amount

Instructions:

1. Count the objects in the box.
2. Circle the correct number.
3. Keep counting!
4. YAY! You're a natural!

Example: Filled Out

Circle the number of items in the box.

0 or 5

2 or 9

3 or 1

Circle the number of items in the box.

fantastic

6 or 8

4 or 9

3 or 7

Circle the number of items in the box.

wonderful

9 or 2

0 or 4

3 or 2

Circle the number of items in the box.

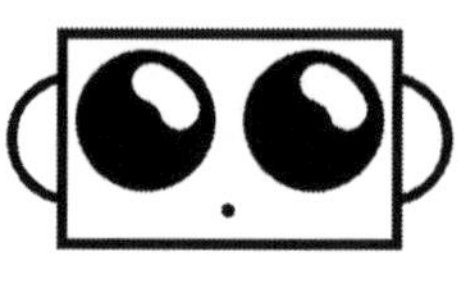
phenomenal

4 or 0

8 or 1

0 or 7

Circle the number of items in the box.

stupendous

9 or 7

4 or 5

1 or 3

Circle the number of items in the box.

wondrous

8 or 4

3 or 0

4 or 3

Circle the number of items in the box.

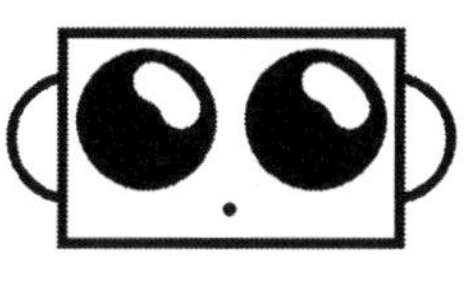

incredible

8 or 0

3 or 9

3 or 1

Circle the number of items in the box.

magnificent

9 or 1

7 or 1

3 or 6

Circle the number of items in the box.

wonderful

0 or 5

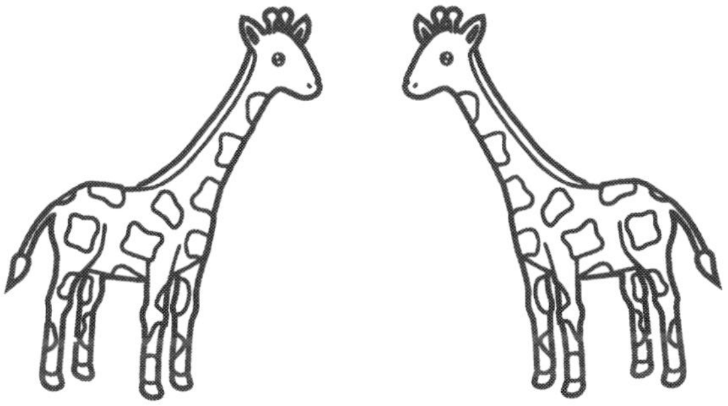

2 or 9

3 or 1

Section 4: How Many?

Instructions:

1. Count how many objects are in a box.
2. Circle the correct number below.
3. Keep going! You're a counting champion!

Example: Filled Out

Circle the number of items in the box.

awesome

1 2 3 4

1 2 3 4

6 7 8 9

0 1 2 3

1 2 3 4

5 6 7 8

Circle the number of items in the box.

0 1 2 3

3 4 5 6

6 7 8 9

6 7 8 9

0 1 2 3

5 6 7 8

Circle the number of items in the box.

remarkable

1 2 3 4

0 1 2 3

6 7 8 9

6 7 8 9

1 2 3 4

4 5 6 7

Circle the number of items in the box.

wonderful

1 2 3 4

3 4 5 6

0 1 2 3

6 7 8 9

3 4 5 6

3 4 5 6

Circle the number of items in the box.

dynamite

3 4 5 6

6 7 8 9

0 1 2 3

0 1 2 3

5 6 7 8

0 1 2 3

Circle the number of items in the box.

terrific

4 5 6 7

2 3 4 5

2 3 4 5

6 7 8 9

0 1 2 3

0 1 2 3

Circle the number of items in the box.

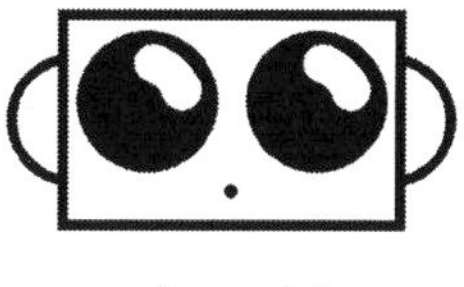

admirable

4 5 6 7

4 5 6 7

0 1 2 3

6 7 8 9

1 2 3 4

0 1 2 3

Circle the number of items in the box.

outstanding

6 7 8 9

6 7 8 9

6 7 8 9

1 2 3 4

3 4 5 6

1 2 3 4

Section 5: Top or Bottom

Instructions:

1. Read the instructions in each box.
2. Circle the correct one.
3. Keep practicing!
4. YAY! You did it!

Example: Filled Out

Top or Bottom!

Color the image the instructions say.

incredible

Circle the top dinosaur.

Circle the top whale.

Circle the bottom T-Bot.

Circle the top owl.

Top or Bottom!

Color the image the instructions say.

wonderful

Circle the bottom duck.

Circle the top basket.

Circle the top monkey.

Circle the bottom rhino.

Top or Bottom!

Color the image the instructions say.

excellent

Circle the bottom turkey.

Circle the top chicken.

Circle the top flamingo.

Circle the bottom tree.

Top or Bottom!

Color the image the instructions say.

sensational

Circle the bottom dog.

Circle the bottom earth.

Circle the top mushroom.

Circle the top hedgehog.

Top or Bottom!

Color the image the instructions say.

super

Circle the top dinosaur.

Circle the bottom house.

Circle the bottom tucan.

Circle the top dolphin.

Top or Bottom!

Color the image the instructions say.

brilliant

Circle the bottom koala.

Circle the bottom planet.

Circle the bottom fox.

Circle the to sea horse.

Top or Bottom!

Color the image the instructions say.

magnificent

Circle the bottom seal.

Circle the top turtle.

Circle the bottom egale.

Circle the top dinosaur.

Top or Bottom!

Color the image the instructions say.

Circle the bottom racoon.

Circle the top T-Bot.

Circle the top octapus.

Circle the bottom rooster.

Section 6: Circle How Many

Instructions:

1. Read how many objects to circle.
2. Count the objects.
3. Circle the correct amount.
4. Your counting skills are amazing!

Example: Filled Out

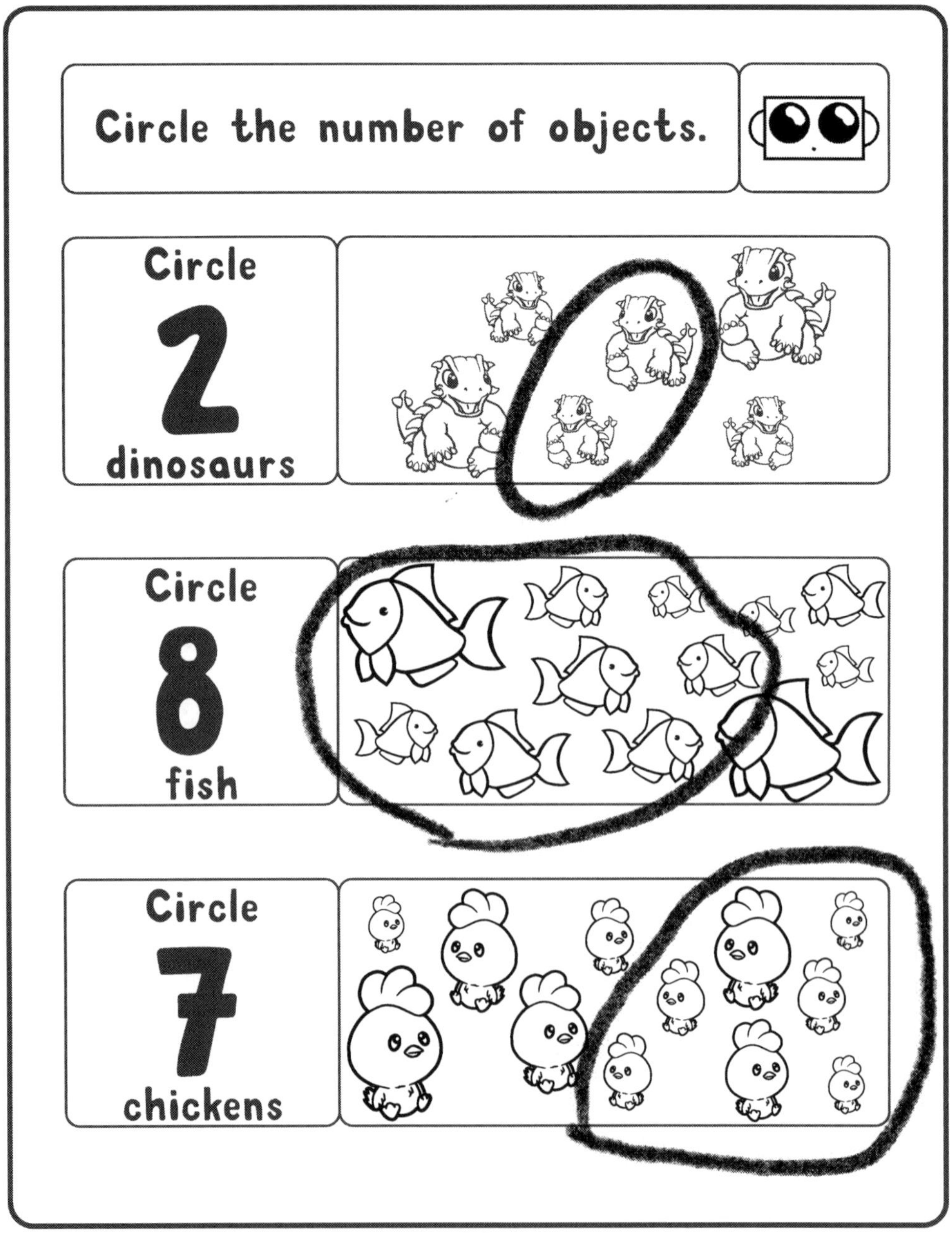

Circle the number of objects.

Circle

4

roosters

Circle

6

chipmunks

Circle

2

planets

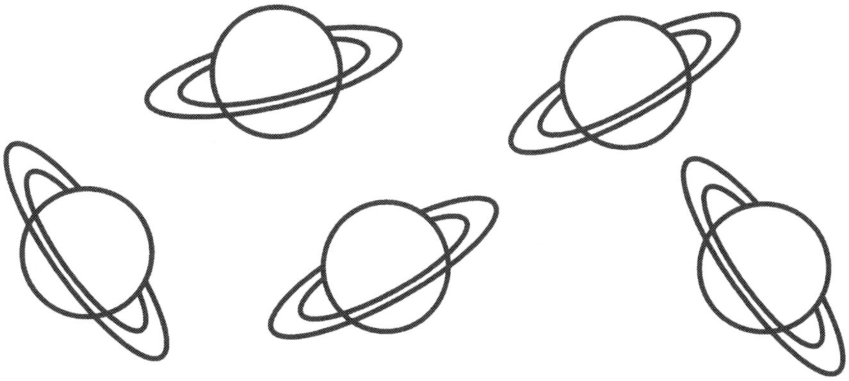

Circle the number of objects.

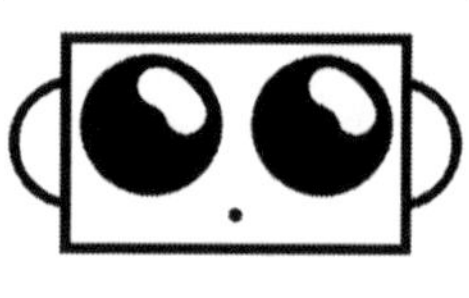
astounding

Circle
3
foxes

Circle
8
mushrooms

Circle
1
earth

Circle the number of objects.

Circle

5

raccoons

Circle

9

dogs

Circle

7

trees

Circle the number of objects.

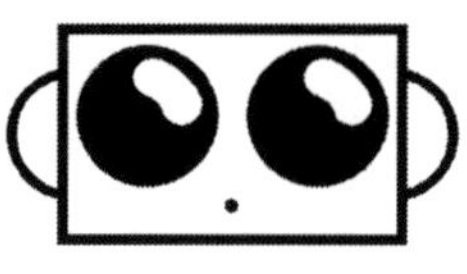

astonishing

Circle

3

dinosaurs

Circle

1

baskets

Circle

7

turtles

Circle the number of objects.

Circle
9
ducks

Circle
1
rhino

Circle
8
crabs

Circle the number of objects.

marvelous

Circle
2
dinosaurs

Circle
6
flamingos

Circle
4
turkeys

Circle the number of objects.

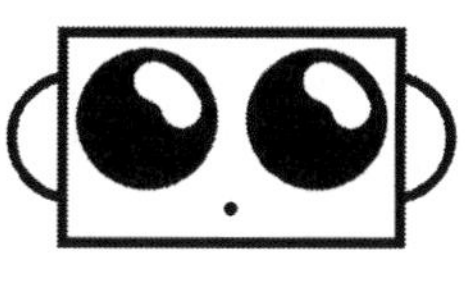

Circle

5

whales

Circle

3

penguins

Circle

1

monkey

Circle the number of objects.

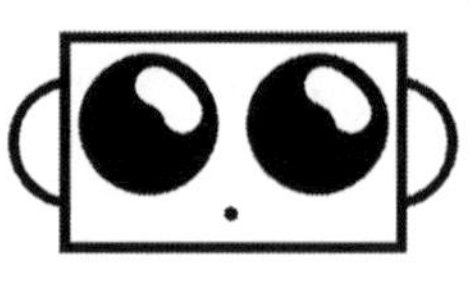

wondrous

Circle

2

dinosaurs

Circle

8

fish

Circle

7

chickens

Section 7: Color the Number

Instructions:

1. Read the instructions on each page.
2. Find each number, then color the number in.
3. Repeat untill you find all of the numbers.
4. You're a coloring and counting genius!

Example: Filled Out

Color all of the 7s

Color all of the 0s

superb

Color all of the 9s

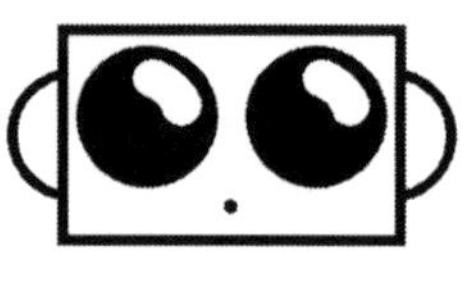

Color all of the 5s

Color all of the 1s

phenomenal

Color all of the 4s

amazing

Color all of the 8s

Color all of the 6s

Section 8: Direction

Instructions:

1. Look at the first image in each section.
2. Then circle the following images that are facing the same direction.
3. YAY! You know your directions so well!

Example: Filled Out

Circle all the objects in the same direction.

Circle all the objects in the same direction.

miraculous

Circle all the objects in the same direction.

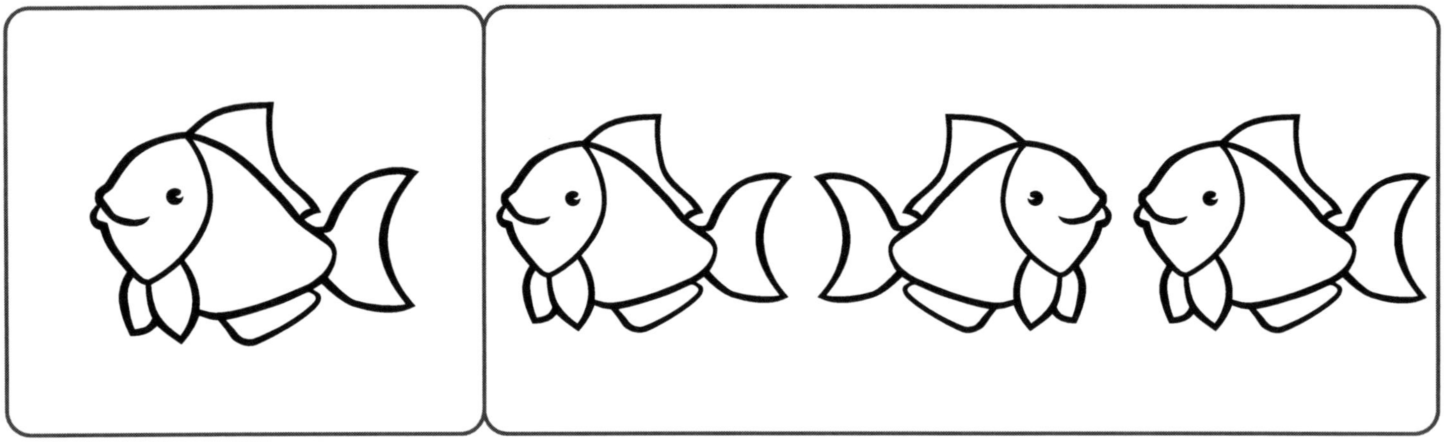

Circle all the objects in the same direction.

fantastic

Circle all the objects in the same direction.

Circle all the objects in the same direction.

remarkable

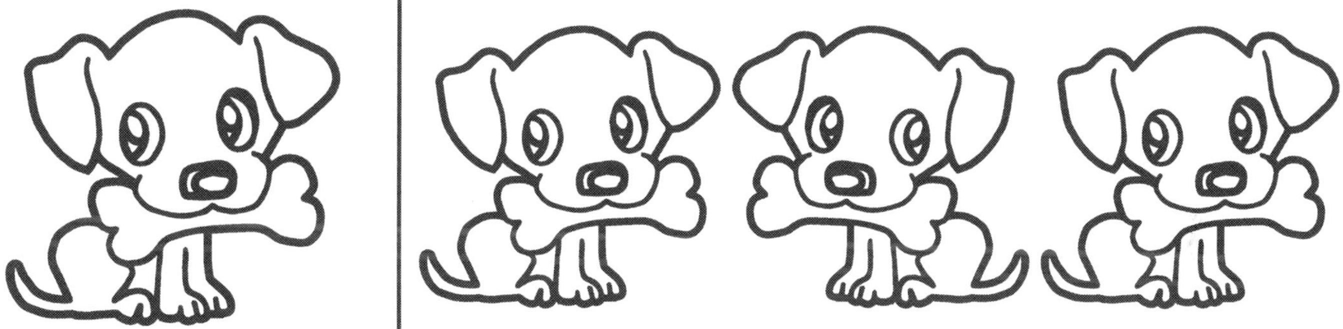

Circle all the objects in the same direction.

Circle all the objects in the same direction.

super

Circle all the objects in the same direction.

Bonus Section 1: Number Sequence

Instructions:

1. In the blank square fill in the missing number.
2. Keep going! You got this!

Example: Filled Out

Write the number that is missing in the box.

4	5	6
7	8	9
1	2	3

Write the number that is missing in the box.

terrific

4		6
1	2	
	7	8

Write the number that is missing in the box.

awesome

3		5
7		9
	3	4

Write the number that is missing in the box.

marvelous

	2	3
7	8	
6		8

Write the number that is missing in the box.

delightful

4 5 ☐

5 6 ☐

1 ☐ 3

Write the number that is missing in the box.

☐ 6 7

☐ 5 6

7 8 ☐

Write the number that is missing in the box.

astounding

2 ☐ 4

6 7 ☐

☐ 2 3

Write the number that is missing in the box.

brilliant

3 4 [ ]

[ ] 2 3

7 [ ] 9

Write the number that is missing in the box.

sensational

☐	5	6
7	8	☐
1	☐	3

Bonus Section 2: Color by Numbers

Instructions:

1. Get your colors! (crayons or colored pencils work best)
2. Color the example crayon the same color as it says.
3. Color each section with the color that matches its number.

Fill in the crayon with the color it says.

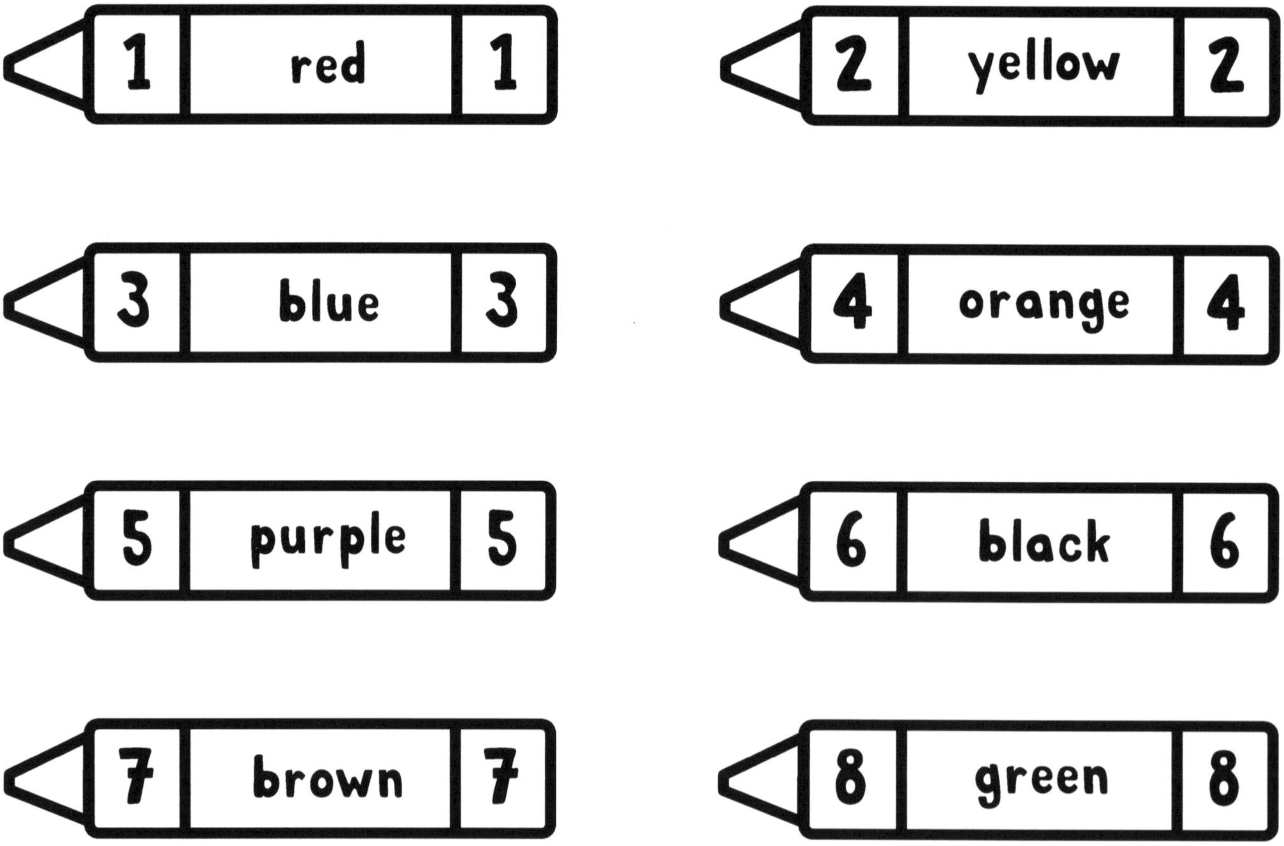

Have fun coloring!

1 red 1

3 blue 3

1 red 1

2 yellow 2

3 blue 3

7 brown 7

5 purple 5

7 brown 7

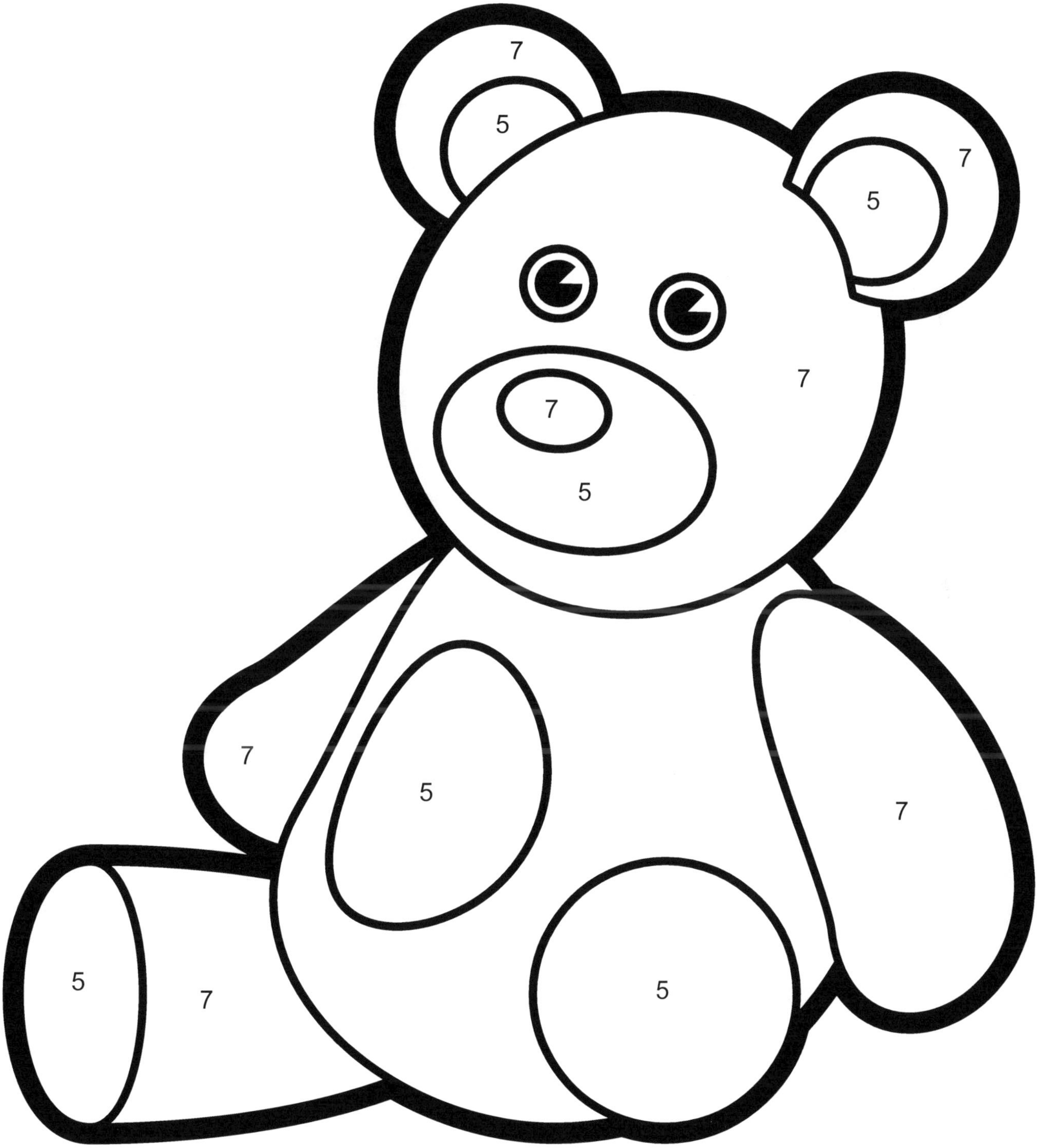

3	blue	3
4	orange	4
5	purple	5
6	black	6

1 red 1

2 yellow 2

4 orange 4

7 brown 7

2 yellow 2

3 blue 3

5 purple 5

4 orange 4

8 green 8

2 yellow 2

5 purple 5

6 black 6

4 orange 4

6 black 6

7 brown 7

1 red 1

2 yellow 2

4 orange 4

8 green 8

1 red 1

2 yellow 2

5 purple 5

8 green 8

Don't forget your FREE goodies!

Scan this QR Code

Congratulations

on completing

T-Bot Learns: Early Math

is awarded to:

____

Presented by:
T-Bot

____
Date

Made in United States
Orlando, FL
07 November 2026

90337534R00059